施道红 / 著　梁玲 / 译
Author: Shi Daohong

In the Knowledge Age
HOW DO WE WORK
智本时代，
我们如何工作

辽宁科学技术出版社

目 录
CONTENTS

环境——创新的触发器
ENVIRONMENT, TRIGGER OF INNOVATION

创新来自"偶遇"
——非正式交流的魅力
Good Air Comes from Plants ... 14

平衡我与我们
Water, King of Medicines ... 22

能让情感连接的空间才是创新的空间
A Space Where Connection Takes Place is the One Where Innovation Happens ... 26

把自然带入办公空间
Bringing Nature Into the Office ... 32

让艺术帮我们创新
Innovate with Art ... 34

无序的曲线大于有序的直线
Chaotic Curves > Orderly Lines ... 38

个性化,才是办公空间正确的打开方式
Individuality is the Way for Correct Office Space Design ... 42

公司文化看得见
CORPORATE CULTURE, MAKE IT VISIBLE

- 办公室里的"破窗效应" — 50
  "Broken Windows" Effect in Offices
- 借小物来承载大文化 — 52
  Small Things for Grand Culture
- 空间即品牌 — 54
  Space is Brand
- 空间布局就是公司文化布局 — 60
  Spatial Layout is Corporate Culture Layout
- 环境塑造互动方式 — 62
  Environment Shapes the Way of Communication
- 会议室也有正能量 — 66
  Meeting Room with Positive Implication
- 办公环境的福利更彰显文化 — 72
  Well-beings of Office Environment is an Expression of the Corporate Culture
- 空间的"无用"之用 — 76
  Make Good Use of the "Useless" Space
- 为未来而设计 — 80
  Design for the Future

越工作越健康
MORE WORK, MORE HEALTHY

- 好空气，种出来
  Good Air Comes from Plants — 88

- 水，百药之王
  Water, King of Medicines — 92

- 吃出百岁的秘诀
  The Eating Secret to Longevity — 98

- 小心！灯光有毒
  Watch Out, the Lamplight is Toxic — 102

- 今天你坐了多久？
  How Long Have You Been Sitting Today? — 106

- 一把好椅子，守护脊椎健康
  A Good Chair, Protection for the Health of Your Spine — 110

- 温湿适宜，才能愉快地工作
  Proper Temperature and Humidity, Enjoyable Working Environment — 114

- 噪声会让工作效率更高？
  Will Nosies Bring Higher Working Efficiency? — 118

- 压力无害，请放心品尝
  Pressure is Harmless, Relax and Enjoy It — 122

- 你的办公室"色"么？
  Is Your Office Decorated with Color? — 126

仅仅在100年前，几乎99%的人还没有工作，朝九晚五更不存在；不过，亨利福特写下了工业时代的第一页，他将汽车的生产流程分为精细的步骤，工人们只需完成某一步，保证没有错误，就可以得到工资。工业时代创造了"工作"这个词。在这本字典里，人=机器。大部分工作的人不需要思考，他们有操作手册，机械地重复工作，只要不犯错误就万事大吉。

只是，突然间，技术改变了一切。机器可以代替人"做"事情，甚至可以代替人"想"事情。我们开始发问——未来的工作中，人有什么竞争力？可能，唯一使我们胜出的，是我们的创造力与我们的人性。未来的工作需要我们释放创造力，释放出我们的潜能，展现出我们之所以为人的一面——我们的爱、希望、勇敢、善良、欲望，甚至是恐惧。未来是创造力的竞争，它将越来越激励，越来越迅速。可悲的是，很多组织还沉溺于旧的工作模式——人＝"机器"。丝毫没有察觉到外面竞争者的敲门声。

Barely 100 years ago, almost 99% of the population did not have a job, let alone a nine-to-five job. But as Henry Fort turned a new page of the Industrial Age by breaking down the process of car production by steps, workers were only responsible for one of the steps and got paid if they made no mistake.

The Industrial Age creates the word "work" that denotes human beings equal machines. Most people working did not have to think, as they have the operation manual. All they had to do was mechanical repetition without making mistakes.

However, all of a sudden, technologies have changed everything. Machines can "do" things for human beings, even "think" for them. We begin to wonder, what edges do human being have in the future work? Perhaps, we can only rely on two things to win the combat with machines: Our creativity and humanity. We need to develop our creativity and potential, and present ourselves as human beings who possess the qualities of love, hope, braveness, kindness, desire, even fear.

Future competition is all about creativity; it's becoming more and more fierce and rapid. Sadly, a large number of organizations are still stuck in the room of the old "human-beings-are-machines" mode, without even noticing knock on the door by their competitors.

这是一个"人 = 创造者",
而不是"人 = 机器"的时代。
我们必须重新定义工作!

This is a "human-beings-are-creators" instead of "human-beings-are-machines" age. We must redefine work.

工作不仅仅是一个地点，或者是我们赚钱的途径，而是一种活法，是我们表达自己的方式，让我们获得成就感，让我们感受到"活着"的状态，甚至给了我们生命的意义！工作让我们越来越健康，越来越聪明，甚至越来越幸福。每个人都可以爱上工作。

这就是我们的愿景，我们希望通过我们的努力，致力于带给面向未来工作的所有人，一个能守护健康、激荡创意、拥有独特文化标识的办公体验。

这就是TOP的意义所在！

Work is not merely about going somewhere, or a way to make money; it is a way of living, a way that we express ourselves, gain a sense of achievement, feel alive and even find the meaning of life!

We become healthier, smarter and even happier. Everyone will love to work.

This is our vision that with our efforts, all office workers in the future can work in an environment that is good for people's health and innovation and that features unique cultural marks.

This is what TOP is about!

普华永道的一项针对于CEO的专业调查显示：61%的CEO表示，创新是其业务的重中之重。

在硅谷，"人与人的互动交流与反馈能激发创新"已然成为一个信仰，不容置疑。我们积累的知识已让我们能够自信地说：创新，不再是偶然，而是必然。

A research of PWC specially designed for CEOs showed that: 61% CEOs said that innovations were the priority among priorities of their businesses. In Silicon Valley, "the interpersonal interaction and feedback can stimulate innovations" has become an undisputable belief.

The knowledge we have accumulated makes us say confidently that: innovation is no longer accidental, but inevitable.

环境——创新的触发器

ENVIRONMENT, TRIGGER OF INNOVATION

**办公室不再是"办公的地点",
而是"创新的工具"。**
Office is no longer "a place of office",
but "an innovative tool".

以前,公司衡量空间效率的关键指标是:每平方米的租金,少有公司衡量空间是否有助于工作业绩。

Zappos的创始人谢家华提出了另外一个指标——"碰撞率"——它指的是每小时在一定面积的人与人交流次数。因为智本时代的工作者因为非正式交流的次数上升,工作业绩会相应提升。 越来越多的公司意识到,空间的租金与员工因为好的空间而带来的收益相比,根本不值得一提。于是他们用设计去打造一个能最大限度促进员工成长、高效工作,切实创新的空间。

In the past, the key indicators for companies to evaluate the space efficiency were the rent per square meter, while few companies cared whether the space would help promote the work performance.

Tony Hsieh, the founder of Zappos, proposed an indicator – "collision rate" which means times of the communications between people in one hour in a certain area. Because in the knowledge age, workers can improve their performance along with more informal communications. More and more companies are aware that compared with the profits brought by employees because of a good working space, the rent is not worth mentioning. Therefore, they design to create a space that can motivate their employees to develop and work efficiently in a maximum and can innovate effectively.

**我们就是要与您探索,
空间设计如何触发创新的。**
We hope to explore with you how
the space design triggers innovation.

创新来自"偶遇"——非正式交流的魅力

INNOVATION COMES FROM AN "ENCOUNTER"—— SPARKS FROM INFORMAL COMMUNICATIONS

皮克斯总部可谓是现代办公室的师祖。1999年，史蒂夫·乔布斯，作为皮克斯的CEO，决定做一个能容下1000人的前所未有的创新空间，并给出了核心思想——这个空间要最大程度地增加偶遇、非正式的交谈、非计划的合作的几率；并且，100年之后，这个办公室还不能过时！

设计师们给乔布斯交了一份满意的答卷，为之后无数拿奖拿到手软的作品奠定了基础。这栋建筑中最为人称道的要数它那大大的中庭，中庭设有一个接待处，有咖啡厅、桌上足球、健身中心，有皮克斯塑造的经典形象玩具，摆放着29个奥斯卡小金人，还有2个能容纳40人的观景房和一个大剧院，还有由乔布斯设计的经典大厕所——全楼只有这一家，人们在洗手的时候被迫见面……

The headquarter of Pixar Animation Studios can be seen as the grandmaster of modern offices. In 1999, Steve Jobs, as the CEO, decided to create an unprecedentedly innovative space that can accommodate 1,000 people, and proposed his core idea -- the space must increase the probability of occasional and informal conversations and unplanned cooperation to the greatest extent; and the office should stay in trend even after 100 years!

The designers came up with a quite satisfactory design proposal, which lays the foundation for their countless works that won countless prizes. In the building, the most commendable space is its big atrium which has a reception, a coffee shop, a table football center and a fitness center and toys of the classic characters created by Pixar. The building displays 29 Oscar statuettes, and includes two viewing rooms that can accommodate 40 people and a large theatre. What's more, it has a classic large toilet designed by Steve Jobs -- the only one in the whole building so that people have to bump into each other when washing hands...

要知道,那年头还是格子间称霸世界的时代,足见乔布斯的远见。 乔布斯在他的传记中写道:"建筑空间也在说话,如果一个建筑物不鼓励交流合作,那就不能获得由高'碰撞率'引发的魔力,从而失去很多的创新,所以我们设计的中庭,就是要让人们走出他们自己的办公室,并与他们平常见不到的人打成一片。"

有趣的是,每一间办公室都有偶遇的机会。 一项针对办公室的研究显示,办公空间的利用率最高值为42%,也就是说,在任何一天的任何时间,你花重金租下来的办公室的一大半都是"空"着的。

Steve Jobs' idea was proven to be quite visionary, since the office cubicles were still predominant at that time. As described in his biography, "architectural space also speaks. If a building does not encourage exchanges and cooperation, it will benefit from the magic triggered by high "collision rate" and will lose a lot of creativity. So the atrium we design is to encourage people to get out of their offices and interact with people they usually don't see."

Interestingly, in every office, there are still occasional opportunities for people to meet. A research on the offices showed that the highest space utilization rate of the offices is 42%. That means over half of the office area that you pay a high rent for is actually "useless".

你的办公室有"偶遇"的非正式交流空间吗？

Does your office have any space for "informal communications"?

那是不是应该搬到一间更小的办公室呢？研究人员继续研究的结果令人吃惊：通过对"空"间的重新改造，让它能增进人与人之间的交流、互动与协助，那么公司的销售业绩及新产品／服务的开发会大大提升，远远超过搬进小办公空间节省下来的成本！

Well, should we move to a smaller office? Researchers continued the research and the result was surprising: through transformation of "empty" spaces, it will promote the communications, interactions and cooperation between people, and then the sales performance and new product / service development will greatly improved, which far outweighs the costs saved by moving into a smaller office!

创新不是一蹴而就，
而是持续不断地修正与进化
Innovation can never be achieved overnight,
but should be constant rectified and evolved.

硅谷，2015年，互联网巨人Facebook落成了新的总部办公大楼，几千个员工挤在一个一英里（约1.609km）长的房间里，散发出一种挤地铁的密度。老大说，这样做不仅仅是让人们合作更便捷，更是利用人与人之间的碰撞，激发更多的创新。

In Silicon Valley in 2015, the Internet giant Facebook completed its new headquarter construction with thousands of employees crowding in a room of a mile (about1.609km) long, a density equal to that in subway. Its boss said that this will not only make people collaborate more conveniently, but will inspire more innovations with collision between people.

那高密度是怎么激发创新的呢？ 想想下棋：在游戏中的任何一点，可能有几步会下得非常巧妙，但是大部分的步子可能并不出彩。 这跟创新是一个道理。 拿印刷机来说，它只可能在活字印刷，纸张和油墨都存在之后才会被发明出来；YouTube，也只有在宽带及视频摄像工具成熟之后，才是一个伟大的想法。回到生命本身，人类出现是在单细胞、多细胞……两栖动物……哺乳动物出现之后才出现的，人不可能直接从鱼类进化而来，而是有一定的次序的。一个点子可以激发近似区域其他的点子，但是隔太远了就不行。 这也是为什么办公室同事之间的距离最好不要超过10米，不然思想就"懒得交流"了。

所以如果你想要一种创新的环境，你就要努力在相邻的领域中创造多样性，同时方便人们去接触这些相邻的想法。

Well, how could the high density stimulate innovation? Think about the chess games: at any point in the games, there may be a few very smart moves, but most of the moves may be not that excellent. This is the same with innovations. Take printing machines as an example, it can only be innovated after the innovation of clay-type printing, paper and ink. YouTube can only become a great idea when broadband and video-recording tools are mature. Returning to the life itself, human beings can only appear after the single cells, multicellular…amphibian animals…and mammals; people cannot directly evolve from fish because there is a certain sequence. An idea can inspire other ideas in approximation fields, but cannot inspire those far way. This is why the distance between colleagues in the offices should be not more than 10 meters, otherwise they will become "lazy to communicate".

Therefore, if you want an innovative environment, you should work hard to create diversity in the adjacent field, and meanwhile, make it convenient for people to get in touch with these neighboring ideas.

工业社会的城市是固态的，
互联网时代的城市应是液态的，
你所在的公司、建筑、园区应该也要是液态的!

A city at the industrial society was in a solid environment,
while the city in the Internet era should stay in a fluid environment.
So should your company, your building and your campus!

MIT 曾经有一个产生无数创新的 20 号楼，曾被称为 20 世纪最好的头脑风暴场所。

20 世纪 50 年代，因为能够用的空间极少，麻省理工学院的 20 号楼挤进了好多不同研究院；有些人来自于生物学专业，有些人则来自于计算机科学，有些是设计，还有做舞蹈的；反正，人员混杂。

MIT once had a No. 20 building which was acclaimed as the best place for brainstorming in the 20th century as numerous innovative ideas were produced here.

In 1950s, ase there were few spaces available for use, many different research institutes crowded into No. 20 building of MIT. Some people majored in biology and some others in computer science; some in design and some others in dance. Anyway, different people mingled with each other.

这些拥有不同学术背景的人经常在一起交流。如果你步行在走廊里，遇到任何人都可以跟他讨论问题，而你可以从不同角度看问题，帮助突破旧思想的局限。

这个环境就是一个液态的环境，是一种介于太多规则与无规则之间的有利于创意的环境。气态太不稳定，固态太稳定，液态则刚刚好。液态是自由的，是开放的，允许争论，允许犯错。思想是要液态的，它可以流动，这样才会促成"信息外溢"，促成创意创新。

建筑空间及城市设计对于促进液态创新非常重要。比如，越发达的城市，包容性越强，来自不同背景的人越多，这些来自不同背景的思想相互连接，相互作用，跨界交流；并且无形中设定了一种"无序"的规则，创新就会繁荣。办公空间也如此，定期邀请不同背景的人来共享空间，举办各种跨界活动，甚至越来越多的公司开放出一小块办公空间，免费提供给其他领域的创业者！

These people with different academic backgrounds often communicated with each other. If you walked in the corridor, you could talk to anyone that you encountered and this enabled you to see issues from a different point of view and break through the limitations of old ideas.

This environment was just a fluid environment, an environment that was between too many rules and no rules and was conducive for creativity. The gaseous environment is too unstable while the solid environment is too stable. So fluid environment is just right. It free and open, allowing arguments and mistakes. Thinking requires a fluid environment which can flow and cause "information spillover" to promote creativity and innovations.

Building space and urban design are very important to promote the fluid innovation. For example, the more developed the cities are, the more tolerant they become to gather more people of different backgrounds. These ideas from different backgrounds interconnect and interact with each other and produce cross-border exchanges. They virtually set up a "disorderly" rule so that innovations can flourish. This is the same with office space, such as regularly inviting people from different backgrounds to share the space, or organizing a variety of cross-border activities. More and more companies even open up a small part of the office space to entrepreneurs from other fields for free.

平衡我与我们
A BALANCE BETWEEN I AND WE

格子农场办公室人人都不是太待见。
The cubicle offices are not much favored by people.

但是开放的办公室就一定好么?
But are open offices surely good?

开放办公是有好处的!它不仅提高了交流沟通的几率;而且增大了办公空间,顿时显得高大上;哦,还有,在新员工进来老员工离开时也不会大费周章,灵活性更高。

但是,办公室开放了就一定是好的么? 未必!

根据 The Sound Agency 的研究:在开放办公环境下的员工效率会暴跌 66%!怎么会这样?

因为你平均每 20 分钟会被打扰一次,根本不能专心。 而且,在众目睽睽之下,你根本没有安全感,怎么专心?

Open offices are good generally. They cannot only improve the chances of communications, but also increase the office space and suddenly make the offices look admirable; Oh, and when the new staff come and the old employees leave, they won't take a lot of energy and have better flexibility.

However, are the open offices surely good? Not necessarily!

According to a research by the Sound Agency: in an open office environment, the efficiency of staff will fall tremendously by 66%! How come?

Because you will be disturbed every 20 minutes on average and cannot concentrate. Moreover, under the watchful eyes of the colleagues, you don't have a sense of security. How can you concentrate?

开放与封闭，一定要有一个优雅的平衡，创意才能真正变成创新，改变世界

There should be an elegant balance between the openness and closeness, so that creativity can truly become innovations to change the world.

皮克斯公司最开始也中了"开放办公"的毒，但是员工普遍反映效率低下，于是后来，他们选择了一个相对密闭的环境；一个工作小组会有一个封闭的U形办公室，里面容纳5～6个人，团队内部既有竞争又有补充，亚马逊的创始人也说，如果2个披萨不能喂饱一个团队的话，那团队也就太大了。

这样，公司将工作空间的尺度重新拉回到人的尺度，同时，员工有更多的不打扰时间去完成自己的创意。

总之，人们花了大量心血研究出来了最理想的工作时间组成：在8小时（甚至更多）的工作时间中，有超过一半的时间处在深度专注状态，大约四分之一的时间在合作，剩下的时间则分配给学习、社交和其他工作。

In the beginning, Pixar was also obsessed with "open office". But generally, employees had a low working efficiency. So later, they chose a relatively closed environment: a working group enjoyed a closed U-shaped office which could accommodate 5-6 people allowing for both competition and complementation within the group. Amazon's founder also said that if two pizzas cannot feed a group, the group is too large.

In this way, the company make the work space back to a human scale, and employees had more undisturbed time to work on their creative ideas. In a word, people spent a lot of efforts in the ideal working time: in eight hours (or even more) of working time, they should deeply concentrate in more than half of the time and cooperate in about a quarter of the time; while the rest of the time should be assigned to learning, social networking and other work.

能让情感连接的空间才是创新的空间

A SPACE WHERE CONNECTION TAKES PLACE IS THE ONE WHERE INNOVATION HAPPENS

Yahoo 在几年前向员工提供了一项可以在家工作的福利，后来管理层又出尔反尔，因为看到别人的公司人气旺旺的，而自己公司内部人与人之间的交流与协助在变少，公司活力在下降，于是硬着头皮决定收回这一项福利。

不过，在网络这么发达、信息如此庞大的今天，为什么我们还是要面对面地协作呢？ 高质量协助很重要的一点是——情感连接。 斯坦福大学做的一项实验，验证情感在人们决策中的重要性。

他们安排了一次向非洲扶贫 NGO 的一个捐赠。 安排了两组人员，给到第一组的是非洲贫困的数据，给到第二组的是一个非洲小姑娘的故事。结果显示拿到故事的捐赠者比拿到数据的多捐了一倍；他们又做了第二次实验，给到另外两组人员故事，但是在故事之前，分别用一个数学题及回忆婴儿的哭声的问题将实验者置于理性与感性之下。 结果，回忆婴儿哭声的捐赠者比做数学题的多捐了一倍，与第一次结果相似。

人的决策很大程度上受到情绪及环境的影响，网络协作很高效，但是员工之间通常都在讨论工作事物，并且物理距离很远，很难有心灵上的沟通。而面对面的沟通会更好地帮助员工之间建立情感，更好地相互协作。

A few years ago, Yahoo allowed its employees to work from home. However, this policy was ended later by the management because they saw the other companies' prosperity with people at the offices, while Yahoo had less and less exchanges and collaboration between employees, which eventually weakened the company's vitality. So the management had to take back the policy.

Yet, why do we still need face-to-face collaboration today when internet has become so powerful with the enormous information so readily available? One important thing about quality collaboration is the emotional connection. In one experiment, Stanford University verified the importance of emotions in decision making.

They arranged a donation to a NGO engaged in poverty alleviation in Africa. They arranged two groups of people, and provided the data of poverty in Africa to the first group and the story of an African girl to the second group. The results showed that those provided with the story donated twice than those provided with the data. Then they did the second experiment, where they provided the story to two other groups. But before providing the story, they put the people into the senses of reason and emotion with a math issue and an issue of recalling baby crying respectively. As a result, those who recalled baby crying donated twice than those who performed mathematical operations, which was similar to that of the first experiment.

People's decisions are affected by emotions and environment to a large extent. Though the collaboration on the Internet has high efficiency, but the employees usually discuss things related to work and they are physically far from each other, so it was very difficult for them to have spiritual communications. While only face-to-face communications can better help employees to establish emotional connections and make better collaboration.

面对面交流，胜在情感连接
The key of face-to-face communications lies in emotional connections.

设计在促进员工协作上也有非常大的作用，比如，澳洲最具创新力的一家公司——AMP，随处可见白板、彩笔、纸，甚至他们的墙都能随写随擦，极大地促进了协作与创新。另外，随处可见的员工照片、涂鸦、食品，私人物品也大大增进了员工之间的情感连接。正是这些，让这家金融公司在金融危机后快速恢复，并在竞争中脱颖而出。

Design also plays a very important role in promoting collaboration among employees. For example, in AMP, Australia's most innovative company, you will see white boards, color pens and paper everywhere, and the employees can even write on the walls and erase what they write at any time. This greatly promote the collaboration and innovations within the company. Besides, the employee photos, scrawl, food and private stuff also greatly enhance the emotional connections among employees. It is exactly such environment that made the financial company recover rapidly from the financial crisis and stand out in the competition.

你敢在公司玩儿么？
很多人都不敢，但是越来越多的公司却鼓励员工去玩。

Dare you play in your company?
Many do not. But more and more companies are encouraging employees to do so.

让我们把工作的乐趣找回来！
Let's get back the fun at work!

比如，皮克斯会举办如纸飞机赛跑、最丑的角色评比的奇葩大赛，你还可以跟各种活体宠物，比如猫、狗，甚至猩猩玩儿！谷歌会鼓励员工拿10%的时间去打沙滩排球、攀岩。总之，去玩儿，而且工资照发不误；工程师们还发明了自己的"流通货币"——Goobles，或交换公司内部资源，或赌博玩。在硅谷，很多公司的会议都是骑着自行车、踩着滑板，一边玩一边开掉的……

智本时代，越来越多的人成为知识工作者，这意味着，我们需要更多的创新。

一项脑部神经研究表明：当你在玩儿的时候，大脑皮质层更活跃，会产生更多的连接，从而生发出更多的想法，同时，你内心抵触或者批评的声音也会消失，创意能更自由地迸发出来。

而如何将"玩"的机制融入办公环境中，让每个人都能在工作中适当地展现出"内心小孩"的一面，是我们努力的方向。

80%的职场人士都在抱怨工作，因为我们日常的工作中没有足够的玩乐，没有激情，也没有爱。如果我们不会玩，不把玩跟工作结合起来，那我们将永远不会爱上工作，工作中也会乏味，缺少创意与激情。工作的反义词不是玩，而是无聊抑郁。

For example, Pixar holds various outlandish contests, such as paper airplane race and the ugliest character. You can also play with a variety of pets there, such as cats, dogs and even orangutans! Google encourages employees to take 10% of the paid working time to play beach volleyball and rock climbing, to name but a few; the engineers even invent their own "currency" - Goobles for exchanging internal resources or gambling. In Silicon Valley, many people hold their meetings when riding bicycles or skateboards. They can both hold the meetings and play at the same time...

In the knowledge age, more and more people become knowledge-based workers, which means that we need more innovations.

A research on brain nerve showed that when you are playing, the brain cortex would become more active and produce more connections, so that more and more ideas would arise; at the same time, your inner conflicting or criticizing voices would disappear, so creative ideas can burst more freely.

The direction of our efforts is to incorporate "play" mechanisms into the office environment and make everyone properly show their "inner child".

80% of employees are complaining about their jobs because of insufficient fun, no passion or love in daily work. If we don't know how to play and don't connect playing with work, then we will never love our work, as a result, work will be always boring and lack creativity and passion. The antonym of work is not playing, but boring and depression.

如何在工作中玩?

虽说没有一个统一标准的范式,但其中非常重要的三点是:

1. 允许失败,接受风险;
2. 有充足的时间与耐心,隔离日常工作;
3. 办公环境中有玩的要素,提醒并激励玩的行为与文化。

How to play at work?

Although there is not a unified standard, has three very important points:

1. to allow failure and accept risks
2. to allow for sufficient time and patience and be isolated from daily work
3. to incorporate playful elements into the working environment and remind and encourage the playing behaviors and culture

"玩上工作"会让员工更有创造性与适应性,更友好,也更具团队精神。

"Playing at work" will make employees more creative, more adaptable and more friendly with more team spirits.

把自然带入办公空间
BRINGING NATURE INTO OFFICES

很多科学研究证明：在医院里，如果病人能通过窗外看到自然风光，那么恢复的速度要明显比看不到自然的病人快。同样，在办公室里，如果我们能被自然所包围，这不仅有助于提高注意力、促进积极的情绪、调节血压等，还能帮助我们从自然中汲取灵感，更具有创造力。谷歌，从某种意义上来说是现代办公一个巨大的试验田，它尽可能地将"大自然元素"带进办公室。谷歌高层说，当引入自然后，我们明显地发现，员工的注意力更集中，他们变得越来越有创造力，效率也更高。

不过，自然可不是那么好请进来的。要知道，大自然是在变化的，温度在早上与晚上会不一样；天空中的云每时每刻都有不同，自然的颜色、纹理也在变化……但在传统的办公室，我们有的是一个恒定的温度与湿度，不变的颜色，相同的纹理……

这么单一的背景下，孕育出像自然那样天才的想法基本是不可能的。所以，我们看到越来越多的办公室，要不就搬到自然中去；要不就把自然请进来。

在 FTA 办公室，我们从中国的园林艺术中寻求灵感，结合大量的科学理论，塑造了新式办公园林。园林既能帮助空气得到优化，也调节了员工在紧张工作时的压力；更为人们提供了一个寻求灵感的好去处。

Many scientific researches showed that in hospitals, if the patients can see the nature through the windows, they will recover obviously faster than those who cannot. Similarly, in offices, if we are surrounded by the nature, it will not only help improve concentration, promote positive emotions and regulate blood pressure, but can also help draw inspirations from the nature and generate more creativity. Google, to some extent, is a huge experimental field of modern offices, because it takes as much the nature into its offices as possible. Google executives said that, after introducing "natural elements", they found employees concentrate much more and become more creative and more efficient.

Well, it is not so easy to bring nature indoors. As nature is always changing: the temperatures in the morning and evening are different; clouds in the sky change at any time; the color and texture of the nature are also changing... However, the conventional offices offer just constant temperature and humidity, the same color and the same texture...

Within such a singular context, it is impossible to produce the genius ideas that resemble the nature. Therefore, we see more and more offices move into the nature or invite the nature indoors.

FTA's office creates a new typology of office garden with reference to the traditional Chinese garden art and the scientific theories. Gardens clean the air, mitigate the people's pressure from intense work, and offer an ideal place of inspirations.

我们希望把自然带入每一个办公空间，让在其中办公的人们有更多的健康、创意与幸福。

We hope to bring the nature into every office and where the office workers can enjoy more health, creativity and well-beings.

让艺术帮我们创新
INNOVATE WITH ART

Steve Jobs 是这样说创新的： 创造力只是连接事物的能力。当你问最有创意的人怎么做到的，他们会感到有点内疚，因为他们只是把一些看到的东西连接起来，并用他们的经验并合成新的东西。之所以能够做到这一点，是因为他们有更多的经验，或者有更多的思考，有更丰富的经历。不幸的是，很多人的背景非常单一，所以没有足够的点去连接，最终他们想到的是非常线性的解决方案，没有从广泛的角度看问题，于是也就没有创新。

Steve Jobs described innovations in such a way: creativity is just the ability to connect things. When you ask the most creative people how to make innovations, they would feel a bit guilty because they just connect things they see and use their experience to synthesize and create new things. The reasons why they can do this are that they have more experiences, more thinking or more backgrounds. Unfortunately, many people have very simple backgrounds without sufficient points to connect things together. Eventually, they just think of linear solutions and do not see things from more perspectives, so there are no innovations.

那问题是：如何快速培养广泛而丰富的经验？

答案便是——艺术

The question is: how to rapidly cultivate broad and rich experiences?

The answer is—— ART

艺术，比如音乐、戏剧和绘画等，它们是灵感的凝结体。欣赏艺术往往更能启发人的思维方式，开放头脑，打破原来的思考框架，勇于寻找解决商业问题的非线性方法。

Art, such as music, drama and painting, are connectors of inspirations. Art appreciation will usually inspire people's way of thinking, open their minds, break the original thinking framework and encourage them to look for nonlinear approach to solve business issues.

美国加州大学的心理学家 Robert Epstein 通过研究发现，当人们被"非寻常"的事物所包围时，或者去一趟博物馆，那么会浮现一些另辟蹊径的想法及解决方案。

Robert Epstein, a psychologist at the University of California in the United States, found in his research that when people are surrounded by "unusual" things or go to the museums, they will have some alternative ideas and solutions.

德意志银行——这家跨国投资银行在 40 个国家的 900 个办公室拥有大约 60000 件艺术品，伦敦总部的接待前台就放置了一件艺术品。作为一家银行，他们竟然设立了艺术总监这个职位，因为他们相信，艺术能帮他们更好地创新。

Deutsche Bank, a transnational investment bank, has about 60,000 pieces of arts in its 900 offices in 40 countries. Even on the reception desk of London headquarter, a piece of art work is provided. As a bank, they surprisingly set up a position of the art director because they believe that art can help them better innovate.

像 Google, Apple, FaceBook 等这些世界上最创新的公司，在他们的办公空间内，随处也能感受到艺术。

Like Google, Apple and Facebook - the most creative companies in the world, the sense of art can be felt everywhere in their offices.

Innovation

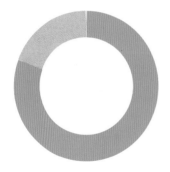 

80% 的自我积累　　　　　　　　　　**20%** 的艺术激发
80% of self-accumulation　　　　　　　20% artistic inspiration

为什么不好好用一下这 20% 呢？
马上就在我们的办公空间内放一些艺术作品，
让它们帮助我们激发创造性的灵感，
帮助企业做到可持续地繁荣。

Why not make a good use of the 20%?
Put some works of art in our offices now, let them help us stimulate
creative inspirations and help enterprises achieve sustainable prosperity.

无序的曲线大于
有序的直线
CHAOTIC CURVES > ORDERLY LINES

有序意味着严谨规范，办公室里的白色、矩形家具，各处可见的90度的直角边等，他们看起来的确很漂亮，整齐干净明亮，经过人为严谨地修饰显得有序极了。但是在这样的环境里面，可能你会感觉到思维并不是那么流畅，甚至，你都不能放松地跟人交流了！

这其实是有原因的，因为在自然中，你很少看到这样"规范"的设计。自然是无序的，它有丰富的曲线，很难看到直线或矩形；它明暗不一、色彩纹理等都有丰富的层次感。

从进化的角度说是成立的，因为无序才有可能孕育繁荣的生命。而这，在一个有序干净无菌的状态中是不可能的。我们对曲线对多样性的偏爱，其实是原始反应。

Order means norms which need to be rigorously conform to. White, rectangular furniture and right angles often found in offices may look really nice, neat, clean and bright, and every appear highly organized after the careful arrangement and trimming. However, within such office environment, you may probably feel that you cannot think freely, or you cannot communicate with others in a relaxing way.

Well, there is a reason. In nature, you hardly see such an "orderly" design. The nature is chaotic, with a lot of curves but few straight lines and rectangles. The nature also has rich layers in light, color and texture.

This is reasonable from an evolutionary point of view, because chaos is the breeding ground of prosperous lives, which is not possible in a clean, orderly and sterile environment. Therefore our preference for curves and variety is actually out of human nature.

多伦多大学的一项科学研究表明：人们更喜欢曲线设计的无序空间，在曲线空间中，他们心情更为愉悦，更能做出积极的决策，员工们也更容易放松警惕，释放自己创新的想法。

A research at the University of Toronto showed that people prefer a chaotic space designed with curves, because in curved space, they feel more pleasant and tend to make more positive decisions and employees are more likely to feel relaxed to express their innovative ideas.

总之,代表无序的曲线设计,让整个办公环境更能激发创新。

In conclusion, design with chaotic curves makes the entire workplace more likely to stimulate innovation.

曲线甚至也能激发人与人之间的交流——创新催化剂。比如,美国人口普查局设计其新总部的时候,就大量用到了"无序"。他们没有像以前一样将座位纵向排得很整齐,而是用了曲线,在设计走道的时候也并不是直线,而是曲线,这样可能让员工更自然地走动,增加"偶遇"的几率。

Curves can even inspire interpersonal communications, which is innovation catalyst. For instance, in design of the new headquarters for the U.S. Census Bureau, a lot of chaotic elements were used. Seats were not neatly arranged into longitudinal rows but along curves. Corridors were designed into curves instead of straight lines, so that employees are more likely to encounter each other when they move more freely.

个性化，才是办公空间正确的打开方式

INDIVIDUALITY IS THE CORRECT WAY FOR OFFICE SPACE DESIGN

每一个人对于办公空间的要求不一样，能激发他们更高效更有创意地工作空间也会不一样。比如，有些人在沙发上灵感迸发，而有些人喜欢在吧台上天马行空；有些人站着时最有灵感，有些人却喜欢独处，一个人专注；有些人喜欢在白板上写呀写；有些人则喜欢小团队的办公室。

所以，如果一个办公室没有弄清楚使用者的需求，并且对于自己的业务目标也模糊不清时，就算把世界上最创新的公司的办公室直接搬过来，估计也没有什么用。

其实，就算是一个员工，在做不同的工作任务时，对于空间的需求也会是不一样的，在专注时，可能倾向于小一点、私密一点的办公空间，而在交流协作时，需要开放一点，能有工具支持的场所，而同时又有非正式交流的空间，放松一下大脑的同时，交流思想、增进情感连接等。

所以，要想有一个最大限度地帮助你的组织目标的办公室，一是一定要从用户角度出发，并结合组织的目标进行空间设计；二是要明白，不同的办公空间适合不同的工作任务，要做好不同空间的融合及平衡。

Everyone has his own requirements on office space and his efficiency and creativity may be inspired by different office spaces. Some are more creative on couches while some are more innovative when they are at a bar. Some are more likely to find inspirational when they are standing, while some are more focused when they are alone. Some prefer writing on whiteboard, while some prefer office for a small group of people.

Therefore, an office design would definitely fail if the requirements of end users are not clear or the purpose is not clearly defined, even though the state-of-the-art office design may be brought in.

In fact, even one single employee has different requirements on space when he is doing different jobs. When his need to focus, he may need a smaller and more private space and when he is collaborating with others, he may need an open space with certain tools. He may also need some informal spaces to relax, exchange thoughts and develop emotional connections.

Therefore, an office design that helps reach organizational targets in the maximum way is definitely the one designed from the angle of end users and with consideration of organizational targets and that provides different spaces for different jobs while spaces are well integrated and balanced.

创新不是一朝一夕的事情，
它是一个慢慢孕育的过程，
一个设计得当的办公空间对于创新来说
是一个非常有效的催化剂。

Innovation is not an over-night thing.
It is a slow process of breeding.
A well designed office space is a very effective catalyst for innovation.

当然，除了形之外，还得有适合创新的企业文化。这就是我们第三章要说到的话题——公司文化，看得见。

Of course, besides architectural form, corporate culture is also important in encouraging innovation, which brings us to the third chapter: Corporate Culture, Make It Visible.

当 AIRBNB 于 2012 年做完 C 轮融资后,他们把金主 Peter Thiel(从 0 到 1 的作者)请到了自己的办公室,问 Peter,你想给我们的最重要的一条建议是什么。Peter 毫不犹豫地说:别搞砸了你们的公司文化。

商界瞬息万变,文化,赋予一个组织鲜明的特色,使其向客户和员工表达的信息更加明确;文化,是所有未来创新的基石;文化,才是能传承 100 年的东西。

When AIRBNB closed their Series C in 2012, they invited Peter Thiel, their investor (author of Zero to One) to their office and asked him what was the single most important piece of advice he had for them. Peter did not hesitate to reply: Don't fuck up the culture.

Business world is fast changing, while culture is the thing that gives an organization a distinct character and helps convey clearer information to its customers and employees. The culture is what creates the foundation for all future innovation and it is the only thing that will endure for 100 years.

公司文化
看 得 见

CORPORATE CULTURE MAKE IT VISIBLE

那如何让大部分的人接受认同企业文化并且切实去行动呢？

这就是我们这一章节要解决的话题——让文化看得见！

How to let most people accept and acknowledge the culture and act as it describes?

This is the subject we need to tackle in this chapter: make the culture visible

那种对于卓越不凡的渴望，你能感同身受吗？存在每个人心中的火种，又如何点燃？

这一切都从一个良好的工作环境开始——全面发展员工的素质能力，不断与时俱进；鼓励大家勇于冒险，敢于创新，不害怕失败，真诚合作、全心投入，并清楚地了解为之奋斗的宏伟目标。

如此一个让员工成长的文化环境才是商业成功的法宝。在一些公司的办公室里，你经常能看到一些核心文化的文字呈现，比如"诚实、开放、创新、进取"等，但可能这些文化唯一存在的地方就是墙上。

为什么会这样？公司文化是非常微妙的，文化是存在于人与人之间的东西，只有当观点、行为、信仰被大多数的员工认可并接受时，才能形成所谓的"文化"。而写在墙上的核心价值观，大多数都是自上而下，由公司的管理层提出，无法被大部分人真正接受，也就形成不了文化。

Can you feel the desire for excellence? How can you ignite the fire in everyone's heart?

It all starts with a nice workplace, a place to fully develop the staff's abilities to keep up with the trend of the world and a place to encourage risk taking, innovation, tolerance of failure, sincere cooperation, commitment, and understanding the ambitious goals to fight for.

Such a culture environment that helps employees grow is the magic for a business to succeed. In some offices, you may see their core values written on the wall, such as honesty, openness, innovation, entrepreneurship, etc. but the wall may be the only place that such culture exists.

Why? Corporate culture is a delicate thing. Culture is something between people, and only when some viewpoints, behaviors and beliefs are commonly acknowledged and accepted by most of the employees, a corporate culture can be built. Core values written on the wall, on the other hand, are mostly top down and provided by the management of a company. They are hardly accepted by most of the employees and are not built into culture.

办公室里的"破窗效应"
"BROKEN WINDOW EFFECT" IN OFFICES

1969 年，美国斯坦福大学心理学家菲利普·辛巴杜 (Philip Zimbardo) 进行了一项实验，他找来两辆一模一样的汽车，把其中的一辆停在加州帕洛阿尔托的中产阶级社区，而另一辆停在相对杂乱的纽约布朗克斯区。停在布朗克斯的那辆，他把车牌摘掉，把顶棚打开，结果当天就被偷走了。而放在帕洛阿尔托的那一辆，一个星期也无人理睬。后来，辛巴杜用锤子把那辆车的玻璃敲了个大洞。结果呢，仅仅过了几个小时，它就不见了。

基于此，学者们提出了"破窗效应"，当一个环境里出现了一些瑕疵，就给人造成一种不和谐的感觉，结果在这种公众麻木不仁的氛围中，坏习惯就会滋生。

In 1969, Philip Zimbardo, a psychologist at Stanford University of the U.S. carried out a research. His team found two cars with almost the same conditions and abandoned one in Palo Alto, a middle class community in California and another one in the relatively chaotic Bronx, New York City. The one in Bronx were removed with license plates and hoods were slightly raised. It was stolen on the very day. The other one in Palo Alto was not even noticed by passersby in a week. However, it was later gone within several hours after the researcher broke the car window with a hammer.

Based on the observation, the term of broken window effect was used to describe the situation when there are environmental flaws, the perception of public disorder will end up in bad habits or behaviors in public callousness.

> 我们塑造了环境，环境也塑造了我们。维护办公环境就是维护公司的文化。
>
> We shape our environment, while environment also shapes us. To defend the office environment is to defend the corporate culture.

办公室的破窗效应也存在，我们要注意那些与公司文化并不相符的空间布局、物品陈设等，及时将"破窗"消除。一个公司崇尚环保节能的企业文化，员工在墙上贴了一个纸条"请节约用电"，下面有人用笔写着"没有人关心"。这个"破窗"被每天路过的人看到，竟然极大地影响了公司员工的参与度，业绩也因为这一个小纸片出现下滑，当纸条移除后，业绩也慢慢回升。

一个好的环境可以成就一个公司，而在一个麻木不仁的环境中，公司也会陷入危机。当这个环境得以维持纯净，我们会感觉到更舒适，有更多动力，全身心投入到公司的使命与愿景中。

Broken window effect also exists in office. We have to pay attention to the spatial layout and furnishings which are not conforming to the corporate culture and eliminate "broken windows" as quickly as possible. In a company that holds to the culture of environmental protection and energy saving, when a note posted on the wall written "please save power" were added with "no one cares", this little "broken window" was seen by passersby everyday and the morale of employees and even performance were greatly affected until the note was removed from the wall.

A good environment makes a good company and a company may fall into crisis with a callous atmosphere. When the environment is maintained flawless, we feel more comfortable, more motivated and more willing to commit ourselves to the mission and vision of the company.

毕加索《公牛》
Picasso's Bulls

借小物来承载大文化
SMALL THINGS FOR GRAND CULTURE

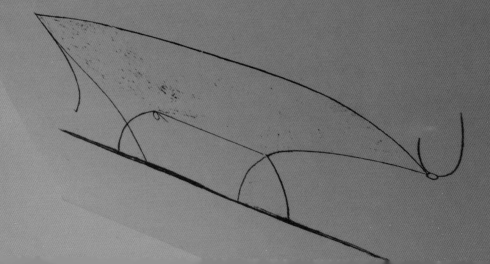

> 要知道，象征物品就是一个个故事，
> 比单纯的"文字"要有力量得多。
> 在合适的地点，你公司内是否有承载文化的象征物？
>
> You know, symbolic objects are like stories and they are more powerful than words.
> Do you have some symbolic objects at the right
> place which convey the culture of your company?

在 FTA 办公室的天才吧，每个人路过时都能看到一个毕加索的公牛，从最初的精细描画到最后寥寥几笔而神形兼备，传递出毕加索这个艺术天才对于"简洁"的理解。这个理念与 FTA 这个建筑设计公司不谋而合，于是，毕加索的公牛做为一个象征物品，放置在人们非正式交流的区域里，每当员工看见它的时候就会受到提醒——简洁，简洁，再简洁。久而久之，内化成整个公司的文化，也通过文化传递到 FTA 设计的每一件作品。

《麻省 - 斯隆管理评论》曾刊载了家具公司特百惠的一个案例，它运用公司内部社交平台的传播力量在短短的三个周末创造了 70 万美金的惊人销量。

听起来像魔术？他们是如何做到的？

很简单，特百惠的首席运营官 Simon Hemus 把写着"创新是氧气"的漫画作品挂在他的办公室门外，每个走过及出入其办公室的同事都能看到，大大地激发了整个公司的创新热情，造就了商业奇迹。

In FTA, there is a Picasso's Bull in the genius bar and everyone who passes by can see it. From the fine drawing at the beginning to the final few divine touches, it conveys this artistic genius's understanding for simplicity, which is identical to FTA's principle. Therefore Picasso's Bull was hung there as a symbol in the informal communication area, reminding employees of our principle of being simple, simple, and simple. Over time, it turns into the company's culture as a whole, which in turn was conveyed to each and every piece of FTA design work.

The MIT-Sloan Management Review once featured a business case of the home product company Tupperware. The company created the amazingly $ 700,000 sales turnover within three weekends by with the help of its internal social networking platform.

Sounds like magic? How did they do it?

It is quite simple. Simon Hemus, the COO of Tupperware Brands Corp posted a cartoon work on his office door, saying "Innovation is oxygen", and everyone who passed by or walked into his office could see it. The passion for innovation is stimulated in the whole company and together they made this business miracle possible.

空间即品牌
SPACE IS BRAND

如果你的办公室可以很轻易地看起来像"邻居家的办公室",别人进来的感觉是"这里我好像曾经来过",你就应该思考思考哪里出了问题。

If your office look just like some other offices, or people who walk in would say "looks like I've been here before", you may have a problem to ponder on.

事实上,品牌不仅仅是LOGO、名片、网站,而更应该延伸到办公空间。品牌应该通过空间深入到员工内心,合作伙伴的内心,办公空间本身就应该是品牌。

这意味着,你的办公空间必须要有一定的独特性,同时与你的品牌身份标示有一定的连贯性;比如,Google的办公室里,你就可以看到Google的LOGO颜色的运用,而不是其他的颜色;在Pinterest(图片分享社区平台),你随处可见大量图片的墙,图片用大别针密密麻麻地订在墙上——这与他们的LOGO是一脉相承的,甚至厕所的男女标志都是用大头针拼出来的。再比如,在FTA,品牌意在表达"用技术与艺术思考未来"的概念,FTA的办公室的室内采用白色,"在这里只有人和思想是有色彩的",从而凸显出人和思想才是企业的灵魂。

这些独特并充分经过思考及设计的细节将空间与品牌融为一体,赋予空间一个身份。通过空间告诉人们,我们是谁,我们相信什么。不管你现在的空间如何,你总是可以将公司的口号、标识、理念、态度、产品和色板融入到环境中,让空间来展示"品牌"。

In fact, brand refers not only logo, name card, website, but also office space. Brand shall be infused into the hearts of employees and partners through the influence of space. Office space itself is part of brand.

This means your office should be unique in certain way and be in consistency with your brand identity. For example, in Google office, the colors of Google Logo are used in the space in stead of other colors. In Pinterest (a photo sharing social network), a lot of photos are pinned on walls, which are consistent with their logo, and even the male and female symbols of the restroom are spelled with pins. In FTA, the brand name intends to convey the philosophy to think about future with technology and art, and white color is greatly used in FTA office interior to express that only people and ideas are colorful here and human beings and thoughts are the life and soul of the firm.

These unique, fully-thought and well-designed details integrate space with brand and give the space identity. Space tells us who we are and what we believe in. No matter how your space is now, you can always integrate your slogan, logo, philosophy, attitude, product and colors of your company into the office environment and let the space speak for brand.

品牌，无处不在
Brand is Everywhere.

英特飞公司被公认为全球最环保的公司之一。在它每一所办公室或公司的前台，都会有一个醒目的公司目标墙——2020、Mission Zero，把环境足迹降低到0。

Interface is recognized as one of the most environmentally friendly companies in the world. In each one of its global offices, there is an eye-catching wall presenting the company's mission: 2020, Mission Zero, reducing environmental footprint to zero.

那个更美好的世界，你的团队是否每天都看得见？
That more beautiful world, can you and your team see it everyday?

意思是：整个公司的运作将不对环境产生负面影响，温室气体的排放降到0，原材料不用一滴原油，不从自然摄取原材料，也不向自然排放任何污染物。这个目标呈现给每一个雇员、供应商、合作伙伴。它是一种文化的外显，唤起人对更美好的未来的向往，激励每个人更努力地为目标奋斗。这种看得见的力量是惊人的，英特飞不但没有因为金融危机而萎缩，反而利润一再提升。

过去，管理者们会认为，公司的商业目标在于"最大化股东的利益""创造超额回报""保持行业领先地位"等。但是现在越来越多的人意识到，这些只是公司在达到商业目标之后的成果而已。而"目标"，关乎于早上是什么把你叫醒？关乎是什么引领着你和你的团队？什么是有意义的？你在两难中会如何抉择？

赚钱，毋庸置疑，很重要。但是，有一个更深层次的东西——目标与意义。找到它，让所有的人看到它，你不仅会赚到更多，生意会蒸蒸日上，而且在你的商业帝国里，人们都会更快乐。

It means the company will eliminate any negative impact of its operation on the environment, reduce greenhouse gas emissions to 0, use not even a drop of crude oil in the production of raw materials, take no raw materials from the nature, and produce no emission of pollutant into the nature. The mission is presented to each employee, supplier and partner. It is an expression of the culture, arousing people's yearning for a better future, motivating everyone to work harder for the mission. The strength of visible mission is amazing. Interface has not shrunk because of financial crisis but its profit has been improved.

Business managers used to believe that the mission of a company is to maximize shareholders' interests or to create excess return, or to maintain its industry-leading position. However, more and more people realized that these are merely byproducts of achieving business missions. Mission is something that wakes you up in the morning, that leads you and your team, that makes sense and that helps you make choices in dilemma.

Making money is, undoubtedly, important, but there are even more important things: mission and meaning. By finding them and making others to see them, you can not only make even more money and achieve even greater success in business, but also deliver happiness to people in your business empire.

空间的布局就是文化的布局。 如果在办公空间内,你能真切地感受到"这个组织在尽力帮我成长",那真是一个很走心的设计!

The spatial layout is the layout of culture. If you can feel that the organization is helping me develop myself in an office space, it must be a very heartful space design.

今天,在很多优秀的企业,公司把最好的空间留给了普通员工,把高管反而放置在一些次要位置。而当人们置身于这样的空间时,立刻会感觉到这个企业对员工的尊重与关爱。 这样的布局,其实是公司文化的一种空间表达方式。

只考虑自己利益的人很难投入自己全部的热爱,倾其所有地去做一件事情,当然最后,也不会收获很多,不会成为领袖,更不可能成就一番大事业。而一个优秀的成功人士最初并不是要让自己成为一个了不起的人,而是在努力地"创造"一些很棒的机会,帮助他人去成功。要知道,一个成功的企业家往往是最后一个得到回报的。

《圣经》里说:"在前的应在后,在后的应在前",就是这个意思。优秀的公司非常懂得这个道理。领袖的职责是帮助更多的人成为领袖,而不是更多的追随者。所以在空间设计中,也考虑到为更多"后来者"提供更多机会,以便成长为新一代的领袖。

Today, many outstanding companies save best space for employees and move executive offices to less important locations. People feel their respect and care for employees as soon as they walk into such space. This layout is the a spatial expression of corporate culture.

Those who think about their own interests can hardly devote themselves into something, and of course have not much return in the end or become leaders of any great career. A successful person may not be someone who wants to be great, but someone who works hard to create good opportunities to help others succeed. A successful entrepreneur is usually the last one to be rewarded.

The Holy Bible says, the last will be first, and the first will be last. An outstanding company knows this saying very well. The leader's work is to help many others become leaders, not followers. Therefore in designing space, we need to think about providing more opportunities for "the last ones" and help them become new leaders.

环境塑造互动方式
ENVIRONMENT SHAPES THE WAY OF INTERACTION

Zappos 的 CEO 谢家华写了一本书《奉上幸福》，阐述了企业文化对于 Zappos 的重要意义。在这本书里，他说，世界上存在的商业问题大都不是因为缺乏创意，事实上，很多企业不差钱、不差人才、不差优质产品，却仍然问题重重。

其病根就在于企业文化——人与人之间的互动方式出了问题。

在 Zappos 的拉斯维加斯的 Downtown 办公室，他们的理念就是牺牲个人空间，最大限度地换取公共空间。 于是，每个人的独立工作区只有 6.5 平方米，是行业标准的 1/3。但是这样一来，人与人之间的距离拉短了， 而且有更多的共同空间，放置超大豆荚工作间、高背沙发、团队大餐桌等，并且有一些诙谐幽默的设计或提醒，鼓励人与人之间轻松的非正式地交流。

Tony Hsieh, CEO of Zappos, has written a book Delivering Happiness, elaborating the significance of corporate culture on Zappos. In this book, he wrote, most of the time, the problem of a business is not lack of creativity. In fact, many businesses do not lack money, nor people, nor quality products, they are still struggling with problems.

The reason lies in corporate culture: they have problems with the way of personal interaction.

In Zappos downtown office in Las Vegas, their idea is to sacrifice personal space for the maximization of public space. Each one's separate work area is only 6.5 square meters, which is one third of the industry standard, but the interpersonal distance is shortened and there are more shared space for large Harwyn pods, high back sofa and large dining table. Humorous design details or reminders are used to encourage informal and free interpersonal communication.

在你的企业里，人与人之间的互动方式是公司文化最理想的反映吗？

In your company, is the way of personal interaction the idealist reflection of corporate culture?

比如，FTA 提倡开放和沟通，希望空间的设计能最大程度地促进沟通，并增加信任；研究表明，人与人之间的连接最主要的影响因素是见面的次数，基于此，在 FTA 的办公空间内，总裁办公室的门是透明的，同时有大片的开放办公场所，每个人都能看见彼此，也增进了人与人之间的连接。

其实，人与人之间的互动方式很大程度上受办公环境的影响。如果你的文化提倡开放平等，但是人与人之间的互动中却有重重大门，需要层层上报才能见上级，那就出了问题。但是如果你的文化人与人之间看重严谨，比如律师事务所，但空间内却非常开放，没有私密谈话的隔间，那也会出现问题。

For another example, FTA advocates open communication and hopes space design can maximize communication and improve interpersonal trust. Researches show that the most important factor affecting interpersonal relation is the number of times they meet each other. Based on this observation, the door to president office in FTA is transparent, and a large open office space is designed to allow people see each other and have better chance for interpersonal connection.

In fact, the interaction between people is largely affected by the office environment. If you say openness and equality are valued in corporate culture but there are actually barriers against personal interaction, and an employee has to report to many layers of managers before he can see the big boss in person, there is a problem. Likewise, if you say your culture values prudence, such as in a law firm, but you have an open office space without separate rooms for private talks, there is also a problem.

706

Meeti

MEETING ROOM WITH POSITIVE IMPLICATION

会议室也有正能量

积极究竟多有力量？研究人员发现，当人们感到消极时，他们的思维会停滞不前，创造力殆尽；相反，积极的情绪和态度会激发人们的创造力与效率，这时候，人们的行为是建设性的。

How powerful is positive attitude? Researches show that when people are feeling negative, their thinking stagnates, and creativity fades away while positive emotion and attitude, on the contrary, stimulate people's creativity and efficiency, and when people are positive, their behaviors are constructive.

其实在办公空间内,有很多空间可以正面积极地激励员工,让他们感受到公司积极文化的力量。其中一个非常重要的地点就是——会议室。

In workplace, a lot of spaces can be positively motivate employees and let them feel the power of company culture. One very important space is the meeting room.

你注意到没，很多会议室就是非常简陋地被命名为"会议室1"、"会议室2"等，而HubSpot这家成功的B2B的营销公司很聪明。他们选出员工们尊敬的人物，以这些人的人名来命名会议室，其中有营销大师，比如Seth Godin, Guy Kawasaki；有商业领袖，比如Steve Jobs, Mark Zuckerberg, Warren Buffett等，这些人有个共同点，就是，他们的言辞与行为能激励员工去工作、去思考，去改变世界。慢慢地，当公司规模扩大时，他们甚至将他们内部的员工、尊敬的合作伙伴的名字作为会议室的名字。这对于员工的积极影响是相当大的，当他们进入到会议室，相互协作时，这些人会时刻激励着他们。

不仅仅是会议室的名称，还有会议室的布局、使用满意度等也在传递着积极的力量。

You may have noticed, many meeting rooms are simply named No.1 meeting room, No. meeting room. HubSpot, a successful B2B company is much smarter. They took a poll of employees on most admired people and named meeting rooms after these admirable figures, including marketing gurus, Seth Godin and Guy Kawasaki and business leaders, Steve Jobs, Mark Zuckerberg and Warrant Buffett. These figures have one thing in common: their thoughts and actions can encourage employees to work, to think and to change the world. Gradually, when HubSpot developed, they named their meeting rooms after their respectful employees and partners. Such positive implication is influential. When they walk into meeting rooms and work together there, these figures become a motivation for them.

Not only names of meeting rooms, but also the layout of meeting rooms and user satisfaction pass on positive implications.

比如FTA大会议室内采用了感应灯和可调节亮度的灯光，这样在使用投影以及不同场景下都可以调到满意的灯光照度，这些细节就是"以人为本"的积极理念。

FTA conference room uses sensor lights and adjustable lights to provide satisfactory brightness when projection is used and in many other occasions. These details are the positive implications of our human-oriented philosophy.

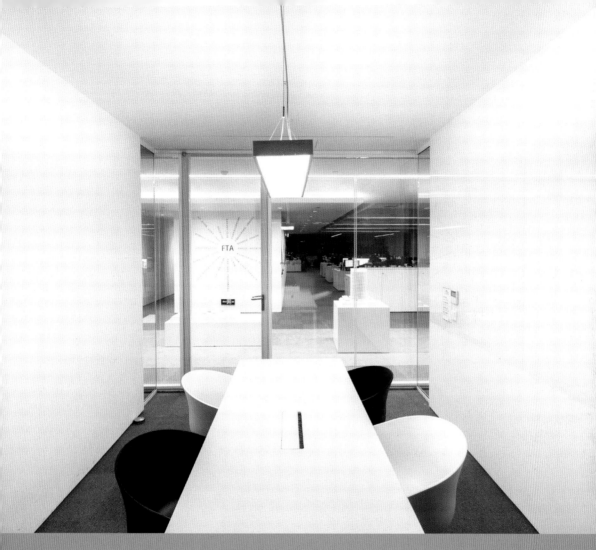

会议室是公司内外交流、使用频繁的场所，是集中体现公司文化的场所。 当与会者走出会议室后，他们是否多了一份积极的力量？

Meeting rooms are heavily used spaces where internal and external communications happen and the place where corporate culture shows itself in a most expressive way. When meeting attendees walk out of the meeting room, are they positively motivated?

办公环境的福利
更彰显文化

WELL-BEINGS OF OFFICE ENVIRONMENT
IS AN EXPRESSION OF
THE CORPORATE CULTURE

可能那里有你的老板、有你的伙伴，你们在一起可以面对面地协作；可能是在办公室，你有更多的灵感，有更多交流的机会；还可能是因为在办公室有福利！

There are your boss and your work partners, you can work face to face with each other there, you might have more inspirations there because you can always talk about your ideas with others, and probably because there are office benefits.

Hey, 你为什么要去办公室?
Hey, why do you go to office?

比如，FTA 的办公室 PM2.5 永远小于 15，室内的空气质量比其他任何地方都要好；再比如，Google 为员工提供 24 小时不间断的餐饮，每个人都可以在长长的自助餐吧免费取食。

In FTA, the office PM2.5 is always lower than 15, and in Google, employees are provided with 24 hour catering services and everyone is able to grab food from food bar for free.

这些小的"福利"甚至会刺激员工每天都从床上"跳起来"来公司工作，更重要的是，它传递给员工公司最看重的部分——文化。FTA 关注员工的健康，相信健康的员工才能创造出健康的公司，所以 FTA 努力通过技术手段，希望每一个员工在这里能安心地工作，当员工走进公司大门，看到显示器上的 PM2.5 数字时，会感觉到安全与关爱，不用担心任何环境带来的伤害。

These little benefits are even the drivers for employees to jump out of bed every morning, and more importantly, they pass on to employees the most important part of the company culture. FTA concern about employees' health, and believe that only healthy people make healthy company. Therefore FTA makes great efforts to help employees work healthily in office with the help of technology. When people walk out of the company and see the number on PM2.5 monitor, they feel safety and care of the company and do not worry about any environmental damage.

那个更美好的世界，
你的团队是否每天都看得见？

That more beautiful world, Can you and your team see it everywhere?

这时，它传递出的是一种健康至上的文化，这个理念会通过浸在这个环境里的员工也感受到，带给合作伙伴及客户。

"福利"并不仅仅是每年发的年终奖，它也可以是空间及办公环境，与公司的文化及价值观达到一定的统一。但要确保的是，员工并不是因为"福利"而待在这里的，而是因为真正相信公司的文化以及愿景。如果你不放心，那可以试一下Zappos的"给钱让你走的策略"。毕竟，文化的纯净性对于一个优秀公司来说是最重要的。

The action demonstrates a philosophy that healthiness comes first and such philosophy will be felt and accepted by employees and through them passed on to partners and customers.

Welfare refers not only the annual bonus, but also the office space and environment and their consistency with corporate culture and value. One thing need to make sure, employees stay in the company not because of these benefits, but because they truly believe in the corporate culture and vision. If you are not sure, try Zappos "pay you to go" strategy. After all, the purity of culture is most important for a good company.

空间的「无用」之用

MAKE GOOD USE OF THE "USELESS" SPACE

在荷兰的乌得勒支火车站（Utrecht railway station）旁边，有一家叫 CDEF Holding BV 的公司，他们的办公室非常开放，不是有开放办公桌椅那种开放，而是把一片区域拿出来，免费给想来这里办公的每一个人，不仅空间是免费的、WIFI 是免费的、咖啡是免费的，还有免费午餐！唯一的要求是：你得用社会资本来支付，即开放的心态与头脑，与你在这里认识的每一位人分享你的知识和才能。

There is a company named CDEF Holding BV next to Utrecht Railway Station in the Netherlands. Their office is open, not in the sense of open office with unpartitioned workstations, but open to the public. They provide a space to anyone who needs a place to work for free, with not only free space, but also free WIFI, free coffee and even free lunch. The only requirement is that you have to pay with your open heart and mind, which they call social capital. That means you have to share your knowledge and abilities with everyone you know in this place.

为什么这样做？
Why do you do this?

因为公司相信可以根据价值观和原则来重新改造市场经济系统，允许个人对社会有更多真正有意义的贡献；同时，也为员工提供更多的灵感，本身的员工也有机会接触到更多元化的人，通过跨界碰撞更多的想法。

Why do they do that? The company believes the market economy system can be reshaped according to values and principles and allow individuals to make more meaningful contributions to the society and this action can provide more inspirations for employees and employees can develop more cross-over ideas through interactions with more diversified groups of people.

在上海FTA的总部，也开辟出来了一片空间，对传统文化中的"庭院"做了改进，打造办公庭院。室内设计中凸显出一个水滴状的生态中庭，员工可以在这里卸下工作的压力，就像在自家庭院散步一样，能够拥有片刻的自然宁静，满足"每一个人心里都有一个院子"的愿望。

回想一下你的办公室？是否有办公功能之外的空间？这个空间它是否让使用者感觉到爱？感觉到激情？感觉到公司的文化？感觉到公司对社会，对环境的一份关怀？

我们每天来到办公室，为了一起协作，把工作做好，也为了成长，为了让自己更强大；更为了激发自己的热情，为这个世界创造出一些美好的事情。

In FTA Shanghai headquarters, there is a space designed into an office courtyard by referencing the traditional chinese courtyard. The teardrop-shaped ecological-friendly courtyard is highlighted in interior design. Employees can shake off their stresses when they stroll around in a home-like atmosphere and enjoy a moment of tranquility in the dream courtyard.

Think about your office. Do you have a space without office functions? Does this space makes users feel loving, passion, the corporate culture and the care of the company for the society and the environment?

We come to office to work together, to work better, to grow, to become stronger and more importantly, to inspire the passion to create some nice things for the world.

功能之外,易见其心
The most delicate part of design lies in the space without function.

为未来而设计
DESIGN FOR FUTURE

你看过人们打冰球吗？如果只是重复它已经滑过的痕迹，那你只有跟着冰球跑；但是如果你预测其轨迹，提前到达它将要去的地方，那你会占得先机。

一个行业里的领头羊企业，始终思考的是企业未来的方向，而不是重复过去的成功。在技术指数发展的时代，商业更新换代的速度也在加快，当然，这并不是说文化也需要快速变化，恰恰相反，文化可能是不变的部分，文化可能是需要坚守的部分。但要注意的是，体现文化的行为可能会发生变化；而承载这个行为的空间也应该相应地进行调整。

比如在过去，很多公司会有一个气派的前台，体现公司文化中好客的一面；随着时代的发展，前台仍然在体现"好客"的一面，但是增加了更多时代的内容，比如让来宾更直观的感受公司的愿景、想法、体验公司的产品等。

Have you seen people playing hockey? If you just skate along the marks the ice hockey had already slid, you can only run after it. You can only get a chance when you predict its trajectory and go to the place before it goes.

A leading company in an industry should always think about its future direction, but not repetition of past successes. In the era of technology indexing development, the business update and replacement are speeding up. Of course it doesn't mean the culture has to change quickly. On the contrary, the culture might be the only thing we need to stick to. It has to be noticed that the behaviors demonstrating the culture might change and the space for such behaviors shall be adjusted accordingly.

For example, in the past, many companies would have a grand reception to reflect the company's culture of hospitality. With time passes by, reception may still be a part of hospitality; it shall be added with more contents of the times, such as allowing visitors have more direct experience of the company's visions, philosophy or products.

一切好的设计都是基于未来，设计现在
All good designs are the designs of now based on future.

更确切地说，未来办公室需要向零售学习。比如 MUJI 的零售店，顾客随处都可以看到 MUJI 在表达"我是谁"；它会向顾客介绍每一个产品背后的故事；MUJI 会开辟一个小角落，放置一些"无用"的艺术展览……总之，顾客来到这个空间后，感受到的不仅是舒适，还有想法与激励。

办公室也没有理由不这样去做——传递自己是谁，吸引客户的眼光，给到员工想法与激情，并鼓励他们去实践。但是，文化的构建是一个不断修正的过程，它需要耐心，需要每个人共同的努力，时常思考办公室应该是什么及为什么的问题，它需要对事业倾注大量的爱！

To be more specific, future office may need to learn from retailers, such as MUJI. In MUJI's retail store, customers can see MUJI is expressing "who I am" everywhere. It tells customers the story behind every product. It arranges a small corner with a little art exhibition of useless pieces. Customers not only feel comfortable in this space, but also are encouraged with new ideas and inspirations.

There is no reason why office cannot do it: tell people who you are, attract customers' attention, bring employees ideas and passion and encourage them to take actions. However, the building of culture is a process of constant modification, and it requires patience, the efforts of everyone inside it, thinking about what an office should be and why and the endless passion for the career.

90% 的职场人士因为工作环境、压力、方式等原因处于亚健康的状态，越来越多的人意识到：人的健康是第一位的,员工不健康的企业是没有竞争力的企业。

90% of the office workers are in sub-health as a result of the working environment, stresses and methods.

More and more people have come to realized that:Health is the most vital. A company whose employees are not healthy is in no way competitive.

越工作
越健康

MORE WORK MORE HEALTHY

这个空间时刻提醒你，保持良好的心境。
在这里办公，你越工作，就会越健康。
The space is a reminder of keeping good mood. The more you work in such a place, the healthier you will be.

想象一个这样的办公室

在这里,

PM2.5 小于 15,空气干净的宛如你置身于海岛上;

光线柔和而明亮,白天让你清醒、兴奋;

温度刚刚好,闻一口,嗯,还有太阳的味道!

在这里,

有着人体工学的家具,让你保持最佳姿势;

累了,触手可及处就是休息空间与健康器材;

还有放心的饮水与食物。

Imagine you are in an office

Here,
PM 2.5 is lower than 15. The clean air makes you feel like you are on a sea island. The light is soft but bright and you are clear-headed and excited. The temperature is just perfect. You take a deep breath. That's the smell of the sun!
Here,
You have the furniture of ergonomic design to keep you in the best position.
When you are tired, leisure space and fitness equipment are within your reach.
You also have safe food and beverage.

好空气，种出来
GOOD AIR COMES FROM PLANTS

现在,请往四周看看有这样的绿植么?
Now, look around, do you have such plants around you?

从1960年代末起,NASA就开始研究"如何在太空中打造一个能适合人类长期生存的环境"。这包括:如何在太空中维持新鲜健康的空气,显然,任务非常困难。空间站内的空气中就有107种挥发性有机化合物(VOC),它们来自建造材料与日常用具,比如甲醛、苯和三氯乙烯等刺激物和潜在的致癌物质。而且因为没有通风设备,情况比在陆地上更糟!科学家们尝试过用更环保节能的材料,但是效果不明显。

这时,一位叫B.C. Wolverton的科学家提出——我们得依附自然的支持系统,好空气,我们种出来!

于是,NASA造了一个类似空间站状态的密闭空间,开始千辛万苦地挑选能净化空气的植物。最终,50种

Since the end of the 1960s, NASA had been working on "how to build an environment that is suitable for long-term survival of human beings in outer space". One of the challenges was to maintain fresh and healthy air in outer space, which was apparently extremely difficult. There were 107 types of volatile organic compounds (VOC) in the air in the space station. These VOCs including stimulants and potential cancerogenic substances such as methanol, benzene, and trichloroethylene came from construction materials and daily-use facilities. Besides, as there was no ventilation facility in the space station, things were even worse than in the Earth! Scientists had tried more environmental friendly and energy efficient materials, but it turned out to be not as effective as they expected.

绿植荣登榜单！在绿植未加入之前，人进入空间会有眼睛灼烧和呼吸困难的症状，而一旦加入绿植，大部分的挥发性有机化合物会被除去，症状消失。1989 年，NASA 就这份研究成果发表了一份报告，引起极大的关注。

在空气污染严重的印度，一个叫 Kamal Meattle 的中年男子开始对德里的空气过敏。医生说他的肺活量已经减少到原来的 70%，这样下去会有生命危险。他必须做点什么，于是，Kamal 利用 NASA 的这份研究成果，在印度理工学院、印度塔塔能源研究所的帮助下，选定了三种基本的绿色植物——散尾葵、虎尾兰和绿萝，保持室内空气的质量。

没想到，几年之后，Kamal 的过敏竟然奇迹般地好了，肺活量也重新回到正常。之后，他联合印度政府，在办公大楼里也做了类似的实验，猜怎么着？

工作效率获得了 20% 的惊人提高；建筑物的能源需求也大幅下降了 15%；在建筑里的人，眼部过敏的情形减少了 52%，呼吸系统问题减少了 34%，头疼症状减少了 24%，肺功能障碍降低了 12%，哮喘减少了 9%！

如果人在这样的建筑空间中待上 10 个小时的话，血氧含量将提升一个百分点的概率达 42%，氧气越多，大脑运转越快，身体也越好。

真的是"越工作，越健康呀"！
Indeed, the more you work,
the healthier you will become!

Then, a scientist named B.C. Wolverton put forward an idea that we must rely on the support of the nature, that is, good air comes from plants. So NASA built a confined space resembling the space station and began to search for plants that purify the air. In the end, 50 plants made the list! Before the plants were introduced, people in the space experienced burning in the eyes and difficulties in breathing. However, once the plants were put into the space, most of the VOCs were eliminated and the syndromes of the people disappeared. In 1989, NASA published a report on the research results which drew widespread attention.

In India where air is heavily polluted, a middle-aged man named Kamal Meattle had been allergic to the air in Delhi. The doctor said that vital capacity had been reduced to 70% of the original capacity, which would eventually become lethal. Something must be done. So with the help of Indian Institute of Technology and Tata Energy Research Institute, Kamal studied the research results published by NASA and selected three plants: chrysalidocarpus lutescens (Madagascar palm), sansevieria (snake plant) and scindapsus aureus (bunting) to maintain the quality of the indoor air.

Unexpectedly, after a few years, Kamal was cured and his vital capacity returned normal. Afterwards, he cooperated with the Indian government to do some similar experiments in office buildings. And guess what?

Working efficiency had been significantly increased by 20% while energy consumption in the building dramatically decreased by 15%. Furthermore, the number of people experienced eye allergy had dropped by 52%, respiratory diseases by 34%, headache by 24%, pulmonary dysfunction by 12% and asthma by 9%!

Working in a space like this for more than 10 hours, there is a probability of 42% that the blood oxygen will be improved by 1%. The more oxygen, the better the brain and the body function.

水，百药之王
WATER, KING OF MEDICINES

《本草纲目》开篇说道："药补不如食补，食补不如水补，水，乃百药之王！"李时珍还花了大量篇幅来讲各种水——露水、明水、夏冰、半天水、井泉水等的疗效，可见水的"崇高地位"。

清朝的乾隆皇帝对于水特别挑剔，他认为：水越轻，水质越好。这个皇帝甚至特意派人制作了一个银斗，去测量各地名泉泉水的重量，结果是："京师玉泉之水斗重一两，济南珍珠泉斗重一两二厘，扬子金山泉斗重一两三厘……"于是，他把整个玉泉山封了起来，凡出自玉泉山的泉水，只能供皇家用。

虽然做法有点霸道，但科学研究证实，水对人的健康有着至关重要的作用。人体65%由水构成，任何食物中的营养都靠水的运输才能被人体细胞吸收；水还是代谢的载体，只有水才能够溶解体内的垃圾、毒素，然后通过汗液、排便方式将这些有害废物排出体外。

At the beginning of Compendium of Materia Medica, the author, Li Shizhen wrote, "Medicine is inferior to food, and food to water. Water, the king of medicines!" He then described in great detail the therapeutic effects of different types of water: Dew, open water, ice in summer, half-day water, well and spring water, fully explained the "elevated status" of water.

Emperor Qianlong in Qing Dynasty was extremely picky about water. He believed that the lighter the water was, the better its quality would be. The Emperor even had someone made a silver cup and measured the weight of renowned spring water in various places. He was told that, the cup filled with water of Beijing's Yu Spring weighted one tael, Jinan's Zhenzhu Spring one tael and two li, and Yangzi's Jinshan Spring one tael and three li. So he ordered the Yu Spring Mountain to be confined and the water of Yu Spring could only be served to the royal family.

Overbearing as he might be, scientific researches have proved that water plays a vital role in maintaining health. 65% of human body is made of water which conveys the nutrition in food to the cells. Meanwhile, water is a carrier of metabolism as only water can dissolve the waste and toxins in the body and discharge them by ways of sweats and defecation.

在广西巴马，平均每10万人就有30.8位百岁寿星，被认定为世界第五个长寿之乡，这里的长寿老人大多数都喝山泉水；在太行山深处也有一个长寿村，村民长期饮用的水源是其长寿的根本原因。长寿村的泉水在水量水温和微量元素含量上常年保持稳定，水源上方还有原始次生林繁茂，其中长满了200多种野生中草药。

在钢筋水泥铸成的办公室内，我们该怎样喝到健康的水呢？世界卫生组织在最新版本的《饮用水水质指南》中指出，饮用水的pH标准范围应在6.5至9.5之间，最好偏碱性（健康人体的pH值在7.35～7.45之间），并且应该是含有益于人体健康的钾、钙、钠、镁、偏硅酸等矿物元素的活水。

In Bama, Guangxi, there are 30.8 seniors aged over 100 every 100,000 people; it is identified the fifth longevity town in the world. Most of the seniors in Bama drink spring water. In Taihang Mountain, there is also a longevity village where the longevity can be attributed to the source of drinking water. The spring water in the longevity village maintains stable in terms of water volume, temperature and microelements, as the source of it is protected by secondary forest that is home to over 200 types of wild herbs.

How can we access to healthy water in armored concrete office buildings? As is pointed out in the latest WHO Guidelines for Drinking-water Quality that the optimum pH required for drinking water is often in the range 6.5–9.5 and water with an alkaline pH is better (the pH of a healthy human body is in the range of 7.35–7.45), containing mineral elements that are beneficial to the health such as potassium, calcium, sodium, and magnesium metasilicate.

那么，你办公室的水是健康的**活水**吗？
So, do you have healthy fresh water in your office?

吃出百岁的秘诀
THE EATING SECRET TO LONGEVITY

对于办公室一族，到底吃什么才能健康长寿？几年前，《国家地理》杂志和美国老龄化研究院的研究人员一起，寻找世界上最长寿的地方。其中一个就在日本冲绳县的古拉格岛。古拉格群岛的中心岛屿上居住着世界上最长寿的女性群体——平均能活87岁，百岁老人随处可见。古拉格岛的人吃了什么？

这些专家仔细分析了人们的饮食习惯，发现：比吃什么更重要的是——如何吃。古拉格岛最重要的一点是：杜绝过度饱食。比如，他们吃饭会用较小的盘子，而且吃饭之前会念一些谚语，诸如：吃饭只吃八成饱。

就这一个小秘诀让人们健康长寿并且充满活力。"八成饱"的饮食习惯也出现在其他的"百岁地区"，比如意大利海岸边的撒丁岛；中国的四川彭山……

而在办公室，八分饱的原则不仅会让人更健康，并且效率更高。如果过度饱食，血液则会过多集中在腹部，这样脑部就得不到充分供氧，效率就会明显降低。

其次，谨记以植物性饮食为主，做到营养来源多样化，奶奶叫不出名字的最好不要放在嘴里。"大道至简"，所有健康都源于知道克制自己的欲望。

For office workers, how to eat healthy? A few years ago, researchers from the National Geographic and the U.S. National Institute on Aging worked together to search for the places where people live the longest. One of the places is Gulag Archipelago, Okinawa Prefecture, Japan. In the central island of Gulag Archipelago lives the female population who lives the longest in the world: an average of 87 years old and seniors over 100 commonly seen. What have they eaten?

The experts carefully analyzed their dietary habits and discovered that how to eat is more important than what to eat. One of the essential features of their dietary habits was no satiation. For example, they eat in small plates and say some proverbs before meals, like: eighty percent full is enough.

This is their little secret to health, longevity and energy. The "eighty-percent full" dietary habit is also favored by some other "longevity regions" such as Sardinia on the beach of Italy and Pengshan in Sichuan, China.

For office workers, adopting such a habit makes them not only healthier but also more efficient. If you eat too much, blood will be gathered in abdomen region and the brain can't get sufficient oxygen supply, which will significantly lower your working efficiency.

Bear in mind that a plant-based diet with diversified sources of nutrition is preferred. You don't want to put anything that your grandma couldn't name in your mouth. The simplest is the most fundamental. Health origins in controlling your own desires.

八分饱，刚刚好！
Eighty percent full is just all right!

小心！灯光有毒
WATCH OUT! THE LAMPLIGHT IS TOXIC

有光的地方，就有希望
Where there is light, there is hope.

根据中国睡眠研究会的调查,在中国,有35%～42%的失眠人群。而其中10%的失眠与灯光有关。而同时,很多办公一族早上总感觉昏昏沉沉没有精力总想睡觉,殊不知,这与灯光也有很大的联系。这到底是怎么回事呢?

首先,我们要知道,我们几千万年进化的结果就是为了适应自然光。所以,不管是什么光,都没有自然光好。自然光线会让我们感到平和幸福;光照不足,质量不高的人工照明可能会引发抑郁症;而太强的人工光照可能会让我们视觉疲劳。并且,自然界在早上与午后存在活跃的蓝光,它让人清醒、增强人的注意力,极大地提高人的工作效率。

人工照明的缺点是:光源要不没有蓝光——人会昏昏沉沉;要不就是蓝光太强——持续到晚上人会太过兴奋而耽误睡眠。更别说自然光可以减轻眼睛疲劳、帮你合成维生素D、省掉照明费等好处了……

但是,很多情况下,对于办公室来说,自然光可能就是奢侈品,怎么办? 你可以利用休息时间到外面去走走,只要15分钟,就会感觉精神焕发! 同时,考虑安装照明效果模仿自然光的照明系统。比如在FTA办公室内安装的灯光系统,很好地模拟了自然光照,让办公有一种在"温和的阳光下"的幸福感觉。加上独特的白色背景的设计,光照强度(电量)比标准少了20%,但是亮度却符合了健康标准。

According to a research of Chinese Sleep Research Society, 35% to 42% of the Chinese population suffers from insomnia; 10% of them involve lamplight. In addition, many office workers feel dizzy and sleepy in the morning. What they never realizes is that their dizziness and sleepiness are to a great extent related to lamplight. Why is that?

First and foremost, we should understand that human beings have been evolving for tens of millions of years to adapt to natural light. In other words, no light is better than natural light. It makes us feel peaceful and happy; but insufficient and low quality artificial lighting may cause depress and overly strong artificial lighting could lead to visual fatigue. Secondly, the natural light in the morning and the afternoon is consists of active blue light that refresh one's mind and heightens one's attention so as to greatly improve one's working efficiency.

Artificial lighting is inferior as the source of light contains either no blue light (causes dizziness and sleepiness) or too much of it (causes over-excitation that interferes with sleep at night). Besides, natural light has more advantages such as easing eye fatigue, assisting in producing Vitamin D and saving lighting cost.

But in most cases, natural light is more like a luxury for office workers. So what to do? You can make good use of your time-outs and walk outside for a while, only for 15 minutes and you will feel refreshed! In the meantime, it is worthwhile to install a illuminating system that imitates the effects of natural lighting. For example, a FTA office lighting system can perfectly imitates natural lighting and people working in the office will feel as if they were working under "the warm sunlight". In addition, its unique design of white background that reduces illumination intensity (power consumption) by 20% but meets the requirements for health.

在世界创新前沿的硅谷，流行着一种病，人称"硅谷病"。这种病主要的症状是：头疼、眼睛干涩、颈椎疼痛、腰肌劳损、手腕损伤、下肢静脉栓塞、肥胖、便秘、痔疮、失眠……大概 60% 的硅谷人都有其中的多数症状，而造成"硅谷病"的病因就是——久坐。

久坐带来的疾病造成了大量的医疗支出、更长的病假时长、工作效率降低等负面影响，以至于坊间戏称——"硅谷最大的敌人其实是久坐！"

俄亥俄州立大学的研究指出，25 岁以上的人，每坐在沙发上或椅子上超过一小时，就会减少 22 钟的生命，其危害是抽烟的两倍！每天若持续长达 6 个小时，久坐不动的话，一生的寿命将会少掉 5 年！世卫组织的报告警示：到 2020 年，全球将有 70% 的疾病是因坐得太久、缺乏运动引起的。

那如何在办公室将运动健康与工作效率完美地结合起来呢？ 在 FTA 的办公室，我们沿窗设计了一排能站着办公的桌子，并且在办公室的墙上粘贴"久坐的危害"，并分享健康运动的体式，提升员工的意识。并且，我们大胆地设计休息及运动区，员工在工作一段时间后，可以选择打打沙袋、做做俯卧撑。让身体恢复活力的同时，头脑更清醒，工作效率更高。

In Silicon Valley where both innovation and a disease called "Silicon Valley Disease" prevail. Such disease features the following syndromes: headache, dry eyes, neck pain, lumbar strain, wrist damage, venous thrombosis in lower extremities, obesity, constipation, haemorrhoids and insomnia. About 60% of the people working in Silicon Valley experience most of the foresaid syndromes and the cause to the Silicon Valley is long-hour sitting.

Diseases caused by long-hour sitting result in so many adverse impacts such as increased medical expenses, longer hours of sick leave, and lowered working efficiency that there is a saying: "The greatest enemy in Silicon Valley is long-hour sitting!"

Researchers from the Ohio State University pointed out that people over 25 years old sitting on the couch or the chair for over an hour will live 22 minutes shorter for every extra hour, twice as damaging as smoking! If one sits for a consecutive 6 hours everyday, one's life will be 5 years shorter! A WHO report warns that by 2020, about 70% of the diseases around the globe will have been caused by long-hour sitting and a lack of exercises.

So how to perfectly combine exercise and working efficiency in office? In the FTA office, we design a line of desks along the window for standing working, and stick posters saying the harms of long-hour sitting on the wall and share healthy ways of exercise with the employees to arouse their awareness of health. In addition, we innovatively design areas for break and exercise so that the employees may have a place to punch sandbags or do push-ups after working for a period of time. In return, they will be rejuvenated and clear-headed; hence improved working efficiency.

一把好椅子，守护脊椎健康

A GOOD CHAIR, PROTECTION FOR THE HEALTH OF YOUR SPINE

在现阶段,办公室大部分时间都是"坐"过来的。澳大利亚的一项研究表明,对于电脑工作者来说,在工作日,除了睡觉,每天直立身体站立运动的时间仅仅为 1 小时 13 分钟,其余的时间都是坐着的……如果你没有坐对,那么不出两三年,腰背和颈椎就会出现不同程度的疼痛,想通过休息一段时间或采用某种偏方来治愈,是不可能的事情。

那如何坐得更舒服,减少腰背和颈椎的压力,保护脊柱呢? 这其实就是人体工学的办公椅要解决的问题!

首先,从生理学上说,人坐着的最佳姿势是坐直了略后仰,上身与下肢大概成 110～130 度,目的是让头枕 + 椅背 + 腰撑共同分担上半身体重。其次,在上身直立的坐姿中,椅子的椅背要能符合脊椎曲度,正好能托住后背的生理弯曲,这有助于打开胸腔,让肺部吸收更多的空气,改善血液流动,运输更多的氧气到大脑,注意力更集中。另外,每个人的体型、体重、身高等都是不一样的,办公用具也应该能根据使用者的偏好进行调整。

当你坐进一把椅子后,你的身体、座椅、思维融为一体了,这才是最好的状态。比起高大上的装修来说,越来越多的员工更看重公司能否提供一把舒适健康的椅子。

Nowadays, most people spend the time in office sitting. A research conducted in Australia indicates that people working in front of the computer, excluding sleeping, spends only 1 hour and 13 minutes on standing and exercising in working days. With such long hours of sitting, if you do it wrong, you would probably experience pains of various extents in your back and neck and there is no way you may recover by taking a break for a while or relying on some folk prescription.

So how to sit more comfortably to reduce the stress on the back and the neck and protect the spine? Let the office chairs of ergonomic design tell you how!

Firstly, from the perspective of physiology, one should be sitting up straight and slightly leaning back with the upper part of the body forming an angle ranging from 110 to 130 degree, so that the headrest, the chair back and the waist support can share the weight of the upper part of the body. Secondly, the chair back should fit the natural curve of the spine and perfectly support the back of the body, which is beneficial to inhaling more air, facilitating blood flow that carry more oxygen to the brain and thereby becoming more concentrated. Meanwhile, as people vary in body type, weight and height, the office furniture should be adjusted according to the preference of the user.

When you sit down on a chair, you body, the chair and your mind are combined as one; the chair brings out the best in you. Compared to fancy decoration, an increased number of employees care more about whether the company can offer them a chair that is comfortable to sit on and good for their health.

坐得好，才是真正的福利！
Good chair is the real benefit!

温湿适宜，
才能愉快地工作

PROPER TEMPERATURE AND HUMIDITY,
ENJOYABLE WORKING ENVIROMENT

温度与湿度对我们的状态有强烈的影响,那在办公室内,可千万别忽视了。

Anyone who has ever experienced rainy season will not deny that temperature and humidity have strong impacts on our states.

20世纪,金宝汤公司(Campbell Soup Company)发现了一个非常有意思的市场洞察。这家公司的产品是汤,他们发现,汤品的销售量与产品本身没有关系,即:汤的口味、其包装设计,甚至是价格与销量没有任何联系。那,什么对销量有关系呢?答案是——天气。

当天气寒冷、潮湿或刮凉风时,人们就会多喝汤。于是,他们决定:在刮风天雨天冷天,买下更多的广告,果真,销售量大大提升了一个水平。

温度与湿度影响着我们的购买决策,同时也影响着我们的工作效率! 美国康奈尔大学的研究人员对这个话题进行了研究,他们发现,当温度相对低的时候(20摄氏度),员工犯错率提高了44%,与温暖舒适的25相比,员工的效率也降低了一半。

冷不仅是不舒服,带来感冒、消化不良等疾病,而且还严重影响工作效率,因为当温度下降,我们消耗能量 保持身体温暖,更少的能量可用于集中注意力、思考和洞察力等智力活动。 而过热会让人烦躁、容易入睡等,更不利于工作效率。

Last century, Campbell Soup Company had a very interesting market insight. The company produce soup. They discovered that the sales volume of the product had nothing to do with the product itself, i.e., the taste, the package design, even the price. Then, what affected the sales volume?

The answer was, the weather.

In chilly, humid or windy days, people tended to drink more soup. So they decided to buy more advertisement in those days, and unsurprisingly, the sales volume was significantly increased.

Temperature and humidity affect not only our decision to buy things, but also our efficiency of working! Researchers from Cornell University studied the subject. The research results showed that when in relatively low temperature (20 celsius degree), employees tends to commit more mistakes (an increase of 44%) and lower the working efficiency (a decrease of 50%) compared to when in 25 celsius degree.

Coldness not only brings discomfort and diseases like cold and indigestion, but also seriously lower the working efficiency as when the temperature drops, the energy we consumed is used to keep our bodies

而同时，湿度应该在40%左右。低于20%的湿度水平会引起皮肤的干燥及不适，也可能引起静电积聚，对办公设备产生负面影响；高于70%的湿度水平可让人胸闷、烦躁，还可能滋生霉菌和真菌。

warm, leaving less of it to support intelligent activities such as concentration, thinking and perception. On the other hand, if the temperature is too high, people become agitated and easier to fall asleep, much more damaging to improvement of working efficiency.

Meanwhile, the humidity shall be around 40%. The humidity lower than 20% will cause dryness and discomfort in the sky and possibly accumulation of static electricity that has adverse impacts on the office equipment; while higher than 70% may cause chest distress and fidgets, and probably facilitate the growth of mold and fungus.

噪声会让工作效率更高？
Will noises bring higher working efficiency?

这是真的！
Surely, it's true!

平常我们都讨厌噪声，WHO 的报告显示，在办公室里，平均的噪声水平是 65dB。从健康角度来说，噪声到了这个程度，人会感到压力增大、荷尔蒙分泌失调、短期记忆退化、注意力与理解力下降、与其他人沟通出现障碍等。

2014 年，Steelcase 与 Ipsos 发现，当旁边有人说话的时候，你的工作效率会降低 66%！每人每天因为噪声污染而损失的时间为 86 分钟！也就是说，每天一个 100 人的办公室，因为噪声会损失 10000 元人民币！

那我们为什么还是坚持噪声会让工作效率增强呢？因为，在一个非常安静的环境中，任何细小的噪声都会使人分心。回忆一下，你在安静的图书馆看书，突然一本书掉落，于是你就分神了。而这种情况在咖啡馆、机场从来不会出现，不是么？

所以，提高工作效率的关键并不是把办公室办成一个图书馆，而是持续产生轻微的、自然的"白色噪声"。白色噪声包括雨声、风声、篝火声、森林流水声、风扇声等，科学家们发现，这些自然动作可以帮助人们专注、提高工作效率。华尔街日报说，适当的背景声音能冲淡人们交谈的声音，让需要专注的人们更能集中注意力，迸发创意。

现在，市面上还有专门的"白色噪声"APP！不说了，我要去感受一下！

Usually, we hate noises. A report released by WHO showed that the average noise level in offices was 65dB. From the health point of view, such a noise level will bring more pressure and cause a hormonal imbalance, so that people's short term memory will degenerate, their concentration and understanding decrease and finally they will have difficulties to communicate with each other.

In 2014, Steelcase and Ipsos found that when someone spoke next to you, your working efficiency would decrease by 66%. The time lost because of noise pollution per person per day was 86 minutes! That is to say, a 100-employee office will lose RMB 10,000 on each day because of the noises!

Well, why do I still insist that noises enhance the working efficiency? Because in a very quiet environment, even the slightest noises will distract people. Recall that when you read in a quiet library, suddenly, a book drops on the ground, and you will be distracted. However, this will never happen in a café or an airport, right?

Therefore, the key to improve working efficiency is not to have a library-like office, but an office with continuously slight and natural "white noises". "White noises" include sound of raindrops, winds, bonfires, forest streams and fans, etc. The scientists found that these natural movements can help people concentrate and improve working efficiency. The Wall Street Journal once reported that appropriate background sound can weaken the voices of people talking and promote people to concentrate, and then creative ideas will burst.

Now, there's a special designed "white noise" APP! No more talking, I'm going to try it out!

在这个智本时代,状态就是竞争力。

所以很多人认为:压力有害健康,我们要把压力降低在一个可控的范围内! 但事实真是这样么?

In the knowledge age, working status becomes the competitiveness.

So a lot of people believe: the pressure is harmful to health and we must reduce the pressure level to a controllable range! However, is it the case?

有一项研究,追踪了3万名美国成人,历时8年,这些美国人被问到两个问题:

1. 去年你感受到了多大压力?

2. 你相信压力有碍健康吗?

在之后的5年中,研究者对比死亡统计数据,发现:

1. 前一年压力颇大的人,并且那些相信压力有碍健康的人,死亡的风险增加了43%。

2. 前一年承受极大压力但不将压力视为有害的人,死亡的风险不会升高,不仅如此,比压力相对较小的参与者都要低,是死亡风险最低的一群人。

One research tracked 30,000 American adults in eight years, all these Americans were asked with two questions:

1. How much pressure do you feel last year?

2. Do you believe that pressure is harmful to your health?

Over the next five years, the researchers compared the death statistics and found that:

1. For these people who bore a lot of pressure in the previous year and believed that pressure was harmful to health, the risk of death increased by 43%.

2. For these people who bore more pressure in the previous year and believed that pressure was harmless to health, the risk of death did not rise and was even lower than that of those who bore relatively less pressure. They are the people with the lowest risks of death.

有人因为压力导致心脏病发，有人却因压力健健康康活到 90 多岁。关键在于看待压力的方式。改变看待压力的方式，更积极地相信你能应付随之而来的压力，生理上的压力反应亦随之改变！

研究发现，更有意思的是，那些花时间关心他人的人完全不受压力影响！这就是我们说的——关爱造就韧性。当你知道，你并非独自战斗，你是带着爱，带着信念去迎接挑战的，压力就会成为你的朋友。

Some people have heart attacks because of pressure, while some others live to be 90 years healthily due to pressure. The key is to how to treat pressure. To change your way of treating pressure and take a more active attitude to believe that you can cope with the consequent pressure will change your physiological stress reaction!

The case research found, what was more interesting was that those who took time to care about others were completely unaffected by pressure! This is what we say – care foster resilience. When you know that you are not fighting alone, but meet the challenges with love and faith, the pressure will become your friend.

总之，压力无害，相信"压力有害"才有害。

In short, the pressure is harmless and
the belief that "pressure is harmful" is harmful.

你可知道，颜色可以引起各种情绪，
进而影响你的工作状态及效率？
Do you know that colors can trigger a variety of emotions and then affect your working status and efficiency?

在得克萨斯大学的一份研究中，研究员安排了三个不同的房间，每个房间涂上红色、白色和水蓝色。并且安排了三组人做文书工作。有些人能"无视"环境的颜色"噪声"，而有些人则会明显地受到颜色的影响。但是，当他们在白色房间里工作时，所有人都会犯更多的错误。大多数工作环境是白色的或者灰色的，这对于工作效率及创意的提高并没有帮助。

那究竟什么颜色才是合适的呢？比如红色，它可提高血压，提高注意力。英属哥伦比亚大学的一项研究发现，如果任务是需要注重细节，那红色是最好的选择。蓝色适合创意工作，蓝色是平静的。它促进沟通、信任和效率，它也能帮助人们以创新的思想开放新思想。在头脑风暴的场所，蓝色是最适合的颜色。绿色能激励创新，促进和谐和平衡。它还可以提高创造性，如果你的办公室鼓励创新，一些绿色将是一个不错的选择。虽然灰色是中性的，但是它是一种缺乏能量的颜色，它让人想睡觉、缺乏信心、甚至抑郁。如果你办公室里有这种颜色，应该用亮色，如红色或黄色来抵消。会议室不要弄成黄色的，因为在黄色背景下，人们更容易激动，发脾气。

我们人眼能分辨出 1000 万种颜色，颜色与我们的满意度、创意、效率有很大的关系。虽然研究成果很重要，但有一点我们得清楚，比颜色更重要的是：使用者对颜色的理解与反应。

In a research by the University of Texas, the researchers prepared three different rooms with each room being painted in red, white and aqua blue respectively. They arranged three groups of people to do paperwork in these rooms. Some people could "ignore" the color "noises" of the environment, while some others were obviously affected by the colors.

Well, what is the suitable color? Such as red, it can raise blood pressure and improve concentration. A research of University of British Columbia found that, if the task is to pay attention to the details, red color is the best choice. Blue color is suitable for creative work, because blue color represents quietness, which can facilitate communication, trust and efficiency. It can also help people initiate new ideas with innovative thinking. In a brainstorming place, blue color is the most suitable color. Green color can stimulate innovation, promote harmony and balance. It can also improve creativity, if your office encourages innovation, some green color will be a good choice. Although grey is neutral, but it is a color lacking energy, which makes people feel sleepy, lack confidence and even become depressed. If there is such a color in your office, please use bright colors, such as red color or yellow color to offset it. The meeting rooms should not be decorated with yellow color because in the yellow background, people are more likely to get excited and lose temper.

Our eyes can distinguish 10 million kinds of colors. Colors make a great difference to our satisfaction, creativity and efficiency. Although the research is very important, but we must be clear with one point: what's more important than colors is the understanding and reaction of users toward colors.

亲爱的读者：

相信你已经快速地看完了这本书。我有写这本书的愿望已经很久，在我刚工作不久，两个朋友因家装有害物质污染，相继离世，这对于两个原本幸福家庭与身边的朋友都是一个沉重的打击。身为建筑设计工作者，从那以后，我一直希望能设计一栋能呵护我们健康的建筑！

20年的积累，作为办公领域的设计专家，我们跟美国的WELL标准合作，与克林顿、比尔盖茨一起投资健康。

现在，我们正在FTA的办公室打造中国第一个"智本时代办公室"的体验馆，我们把它叫做TOP(The Office Project)——办公室计划，它会体现在三个层面：

1. 健康；
2. 创新及效率；
3. 公司文化。

恰好与这本书的三个章节相呼应。

如果您也想加入这份事业，就请联系我们！
If you want to join this project, please contact us!

Email: shidh@ftaarch.com

Dear readers,

I believe you have quickly finished reading the book. I wanted to write this book for a long time. Not long after I started to work, two friends of mine passed away because of harmful substances from home decoration. These were heavy blow for two otherwise happy families and many friends of mine. As an architect, I always hope I can design a building that could protect our health.

As an architect office specialized in designing workplace for over 20 years, we work with WELL Building Standard and invest in health together with Bill Clinton and Bill Gates.

Now, we are building FTA office into the first experience center of "Office in the Knowledge Age" in China. We call it TOP (The Office Project). It will be unique in three ways:

1. Health (work well);
2. Innovation and Efficiency (work smart);
3. Corporate Culture (work with love), which happen to be coincident with three chapters of this book

这个世界存在的问题太大太多太紧急，
我们仍然要不顾一切地去奋斗，
努力让这一切实现。
我们需要建设性的思考、勇敢的行动。
如果我们想要一个更好的世界，
我们必须合作，
我们必须重新定义"增长"，
我们相信智慧存在于每一个人，
相信每个人心中都有善良与同情，
我们只是需要发现它，
并激励更多的人同行。

Though there are so many, so big and so urgent problems in this world, we need to fight against all odds to realize what we work so hard for. We need to think constructively and act bravely. If we want a better world, we have to work together and we have to redefine "growth". We believe wisdom is with everyone and everyone is kind and compassionate. All we need to do is to discover them and encourage more people to work along with us.

全球最顶尖的企业是如何通过空间设计来关照员工、激发创意和树立文化的?

How do the top companies care about their employees, stimulate innovation and develop corporate culture by spatial design?

感谢玛祖铭立为本书提供图片

玛祖铭立上海旗舰店
MATSU Shanghai Flagship Showroom
Shanghai, China

T +86-(0)21-6048 8001
F +86-(0)21-6431 6466

No. 686 Zhaojiabang Road, Xuhui District, Shanghai, China 200030
中国 上海市徐汇区肇嘉浜路 686 号 邮编 200030

In the Knowledge Age
HOW DO WE WORK
智本时代
我们如何工作

施道红 / 著　梁玲 / 译
Author: Shi Daohong

FTA Group GmbH
Shanghai, China

T   +86-(0)21-6590 9056 / 6590 0015
F   +86-(0)21-6590 9049

7.F Baoland Plaza B, 588 Dalian Road, Shanghai,China 200082
中国 上海市大连路 588 号宝地广场 B 座 7F   邮编 200082

www.ftaarch.com

Copyright © FTA Group GmbH. All rights reserved.

Follow us in

Wechat:
fta_group

图书在版编目（CIP）数据

智本时代我们如何工作：汉、英 / 施道红著；梁玲译 . -- 沈阳：辽宁科学技术出版社，2016.7
ISBN 978-7-5381-9871-3

Ⅰ . ①智… Ⅱ . ①施… ②梁… Ⅲ . ①办公室 – 室内装饰设计 – 汉、英 Ⅳ . ① TU238
中国版本图书馆 CIP 数据核字 (2016) 第 158096 号

出版发行：辽宁科学技术出版社
（地址：沈阳市和平区十一纬路 25 号邮编：110003）
印 刷 者：辽宁奥美雅印刷有限公司
经 销 者：各地新华书店
幅面尺寸：170mm×180mm
印　　张：6
插　　页：4
字　　数：50 千字
出版时间：2016 年 7 月第 1 版
印刷时间：2016 年 7 月第 1 次印刷
责任编辑：杜丙旭
封面设计：徐　晨
版式设计：麻亚灵
责任校对：周　文

书　　号：ISBN 978-7-5381-9871-3
定　　价：128.00 元
联系电话：024-23284360
邮购热线：024-23284502